绿色生活　从心出发

Green living blooms from the heart

我的绿色低碳生活

Little Green Ideas:
A Low-Carbon Lifestyle Picture Book

宋明霞 著
陈慧玲 绘图 廉子健 译

ACADEMIC ADVISORS

学术顾问

杜祥琬　中国工程院院士、中国工程院原副院长、国家能源咨询专家委员会副主任、
　　　　国家气候变化专家委员会顾问
潘家华　中国社会科学院学部委员、国家气候变化专家委员会副主任
舒印彪　中国工程院院士、中国电机工程学会理事长、国际电工委员会（IEC）第 36 届主席
江　亿　中国工程院院士、国家气候变化专家委员会成员、清华大学教授
孙宝国　中国工程院院士、中国食品科学技术学会理事长、北京工商大学国酒研究院院长
刘　坚　全国农业科技创业创新联盟主席，全国乡村文化创新联盟主席，农业部原副部长
张来武　中国软科学研究会理事长、复旦大学六次产业研究院院长、科技部原副部长
仇保兴　教授、国际欧亚科学院院士、住建部原副部长、国务院原参事
傅成玉　中国国家碳中和科技委员会委员、COP28 阿联酋国际顾问委员会委员、
　　　　碳中和国际研究院创始院长
朱宏任　中国企业联合会、中国企业家协会党委书记、常务副会长兼秘书长
杨　昆　中国电力企业联合会党委书记、常务副理事长
王思强　国际金融论坛（IFF）能源转型发展委员会主席、中国电力建设企业协会会长
闫世东　生态环境部宣传教育中心党委书记、主任
涂瑞和　联合国环境规划署驻华代表
杜少中　中华环保联合会荣誉副主席
吴绮敏　人民日报数字传播有限公司董事长、总经理
徐晋涛　北京大学国家发展研究院教授
柳冠中　清华大学教授、中国工业设计协会荣誉副会长
刘素文　中国自行车协会理事长
周建平　中国电力建设集团有限公司首席技术专家、原总工程师
陶　岚　中华环保联合会常务理事、碳普惠专委会执行主任
李长栓　北京外国语大学高级翻译学院教授、副院长

Change Begins with You and Me

While we have reaped the prosperity and progress of the Industrial Revolution, we now face the profound challenges of accelerating global climate change, environmental pollution, and ecological degradation. As a researcher dedicated to energy strategy, environmental protection, and climate change, I deeply feel the responsibility—yet I am also buoyed by hope.

China is committed to achieving the goal of carbon neutrality. The path to a comprehensive green transformation in economic and social spheres requires a dual approach: promoting both green, low-carbon production and sustainable, low-carbon lifestyles. It calls on each citizen to take action—decisive, long-term action.

The publication of *Little Green Ideas: A Low-Carbon Lifestyle Picture Book* is particularly timely. Ms. Song Mingxia, with her profound expertise and keen insights, has artfully transformed the concepts of green and low-carbon living into vivid and relatable scenarios through this illustrated book. Her work radiates warmth, simplicity, and charm, while offering thought-provoking ideas and wide-reaching appeal.

This book is not just a repository of knowledge; it is a rallying cry. It invites each of us to begin with small acts in our daily lives, reminding us that even the smallest gestures can carry the seeds of great change. When each individual embraces a green and low-carbon way of living, we can create a better world together.

I sincerely hope that *Little Green Ideas: A Low-Carbon Lifestyle Picture Book* becomes a trusted guide for readers, especially young people, inspiring their passion for discovering and practicing green living. We believe that change begins with you and me, and that the future will grow brighter through our commitment to a greener path.

Du Xiangwan

September 2024

序

改变从你我开始

我们享受了工业革命带来的繁荣与进步，也不得不面对全球气候变化加剧、环境污染和生态退化带来的严峻挑战。作为一位长期关注能源战略、环境保护与气候变化的科研工作者，我深感责任重大，同时也满怀希望。

中国正以坚定的决心推动实现“碳达峰碳中和”目标。加快经济社会发展全面绿色转型，需要绿色低碳生产和绿色低碳生活双轮驱动，需要每个公民付诸行动——果敢的行动、长期的行动。

《我的绿色低碳生活》一书的出版恰逢其时。作者宋明霞女士以其深厚的专业功底和敏锐的洞察力，通过生动形象的绘本形式，巧妙地将绿色低碳理念转化为具体的生活场景，亲切、朴素而温暖，兼具趣味性、思想性和传播性。

本书是知识的传播者，更是行动的号召者。它倡导我们每个人从生活小事做起，看似微不足道的举动，实则蕴含着巨大的改变力量。当每个人自觉践行绿色低碳生活，我们的地球家园将会变得更加美好。

在此，我衷心希望《我的绿色低碳生活》能够成为广大读者，特别是青少年朋友的良师益友，激发大家探索和践行绿色低碳生活的热情。我们相信：改变，从你我开始；未来，因绿色而美好。

杜祥琬

2024 年 9 月

Everyone is a Micro-unit for Zero Emission

The scorching summer of 2024 reminded us of the urgent threat that climate change poses to all of humanity. To confront this global challenge, we must accelerate the transition to renewable energy, gradually move away from our dependence on fossil fuels, and embark on an unprecedented journey towards a future of zero-carbon emissions. This journey represents not only a profound transformation in how we produce energy but also a fundamental shift in our lifestyles and mindsets.

I have known Mingxia for many years, and together, we have witnessed the remarkable transformation of China's energy landscape. We have often delved into the challenge of climate change and explored pathways to a green, low-carbon lifestyle. In a world that rushes forward, Mingxia took the time—three and a half years—to carefully craft this picture book, a work that reflects her deep sense of responsibility and creative spirit. The book draws from the perspective of an everyday Chinese citizen, presenting a narrative that is simple yet profoundly warm. It calls for a collective embrace of green and low-carbon actions, urging a shift towards green consumption that can, in turn, drive green production—rich in thought, beautifully illustrated, and brimming with meaning.

As we move toward a zero-carbon future, each of us becomes a crucial part of this larger vision—a zero-carbon micro unit. I am confident that the publication of *Little Green Ideas: A Low-Carbon Lifestyle Picture Book* will not only showcase the dedication of 1.4 billion Chinese people in addressing global climate change but also inspire enthusiasm for green, low-carbon living far beyond our borders and foster a more sustainable future worldwide.

Pan Jiahua

September 2024

序

每个人都是零碳微单元

我们在高温炙烤中度过了 2024 年夏季，切身感受到气候变化给人类带来的严重威胁。为了应对全球气候变化，我们不得不加快发展可再生能源，逐步减少对化石能源的依赖，开启一次前所未有的能源转型，探索全面零碳排放的未来。这一过程，不仅是人类生产方式的深刻变革，更是人们生活方式与思维模式的根本性转变。

明霞与我相识多年，共同见证了中国能源发展和转型的波澜壮阔，也曾多次共同探索应对气候变化、践行绿色低碳生活的实践路径。在快节奏的时代，明霞不惜花三年半时间“练慢工”，《我的绿色低碳生活》一书背后是她强烈的责任心和创新精神。绘本从“我”，即普通中国人的生活视角切入，朴素而温暖。呼唤全民绿色低碳行动，提振绿色消费，以绿色消费倒逼绿色生产，以绿色消费推动绿色生产，绘本思深图美、内涵丰富、意义深远。

迈向全面零碳的未来，我们每个人都是零碳微单元。我相信《我的绿色低碳生活》的出版，将向全球读者展示十四亿中国人为应对全球气候变化所做的努力，激发社会各界对绿色低碳生活的认同和参与热情，对全球推广绿色低碳生活方式发挥积极作用。

2024 年 9 月

A Greener World, A Brighter Future

Embracing a life of simplicity, moderation, and low-carbon living is essential to China's path of sustainable progress and the fulfillment of the United Nations' Sustainable Development Goals. As we ride the wave of collective effort toward our dual carbon objectives, the publication of *Little Green Ideas: A Low-Carbon Lifestyle Picture Book* is a welcome and timely addition. This moment calls for wisdom and creativity, and Ms. Song Mingxia has answered that call. As a respected member of the Carbon Inclusion Cooperation Network Expert Committee at the Ministry of Ecology and Environment's Public Communication Center, she has devoted herself to exploring China's energy, power, and low-carbon innovations. With an unwavering commitment to fostering a green lifestyle, she has crafted this picture book with both care and clarity, blending strategic foresight with an expansive vision. The work gathers everyday green actions and low-carbon habits, presenting them like a radiant string of pearls, each action shining with purpose and possibility.

This book serves as a bridge, linking our daily routines to the Earth's future. We believe that when individual choices come together, they form a powerful current capable of reshaping the world. Each of us stands as a guardian of this shared, vibrant life on our planet.

For years, the Ministry of Ecology and Environment's Public Communication Center has been devoted to spreading the ethos of green, low-carbon living. Now, let us join hands and turn intention into action—embedding green values in our hearts, making low-carbon choices a part of our everyday lives, and spreading the beauty of a sustainable lifestyle to every corner of the world.

Yan Shidong

September 2024

世界因绿色而美好

简约适度、绿色低碳的生活方式，是实现中华民族永续发展和推动联合国可持续发展目标实现的重要途径。

在全社会共同努力奔赴“双碳”目标的绿色浪潮中，我们欣喜地看到《我的绿色低碳生活》一书的出版。作者宋明霞女士是生态环境部宣传教育中心碳普惠合作网络专家委员会成员，长期致力于我国能源电力和绿色低碳研究。为推广绿色低碳生活方式，她反复打磨，从战略的高度、以开放的视角精心编写了这部绘本。全书以清新隽永、轻松活泼的风格，将生活中的绿色低碳行为一一拾起，串成一条美丽的珠链。

《我的绿色低碳生活》是一座桥梁，连接着我们每个人的日常生活与地球的未来。我们相信，我们都是地球生命共同体的守护者，单个人的选择汇聚起来，就能形成改变世界的力量。

生态环境部宣传教育中心一直致力于推广绿色低碳理念。让我们携手共进，以实际行动践行绿色低碳生活，让绿色理念深入人心，让低碳生活形成风尚，让绿色低碳美学风行世界。

闫世东

2024 年 9 月

Bringing a Verdant China to the World

This picture book is a tribute to nature, a celebration of a green, low-carbon lifestyle—a book that invites both adults and children to explore its message with ease. Each page paints a scene, vivid with color and emotion, embodying our deep commitment to sustainable living.

It is not a lofty dream, but a heartfelt appreciation for the beauty of everyday life—not a deluge, but a gentle stream that nourishes quietly. It favors moderation over extravagance, imparting the core principles of green living.

Through the subtle details of daily life, this book invites you to uncover the quiet beauty of a low-carbon lifestyle. It fosters an appreciation for the aesthetics of sustainability, sparks new ideas for green living, and nurtures a culture of environmental consciousness.

The moment you open this book, we embark on a journey together to celebrate the harmony between nature and life. As you turn its pages, you will discover the immense power of small, everyday actions in preserving our shared home.

Act now—drive less and choose public transportation, order thoughtfully to avoid food waste, sort your garbage, and conserve every watt of electricity. The greenest energy is the energy saved, and the most sustainable resources are those preserved.

The most profound change comes when every family and every individual takes the lead. With 1.4 billion hearts and hands united, we can present to the world a China where blue skies and green landscapes stretch unbroken to the horizon.

Song Mingxia

September 2024

自序

给世界一个碧色连天的中国

这是一本敬畏自然、倡导绿色低碳生活的绘本；

这是一本大人和孩子都能轻松阅读的绘本。

每一页都是一个小故事，是我们用心捧出来的绿色小景；

每一页都浓墨重彩，是我们对绿色生活方式的热烈追求。

不是奢华的梦想，而是对平凡生活的真挚赞美；

不是倾盆大雨的漫灌，而是涓涓细流的浸润。

提倡使用，不鼓励占有，践行绿色低碳生活方式。

传递悄然的生活细节，在平凡中感受低碳生活的美好；

倡导绿色低碳美学，表达点滴绿色小心思；

激发绿色低碳生活创新灵感，培养绿色文化气质。

您翻开绘本那一刻，我们就一起漫游于自然与生命的交响之旅；

您对话绘本那时起，我们就携手开掘小微行动之于生态家园的巨大力量。

最好的开始就是行动。少开私家车多乘公共交通、适量点餐杜绝浪费、垃圾分类、节约每一度电……

节约下来的能源最绿色，节约下来的资源最绿色。

最有力的行动莫过于每个家庭、每个人身体力行。

14 亿人付诸行动，我们给世界一个碧色连天的中国！

2024 年 9 月

目录 Contents

我们正在面对严峻的现实：温室气体排放导致地球气候变暖、极端天气频发，环境急剧恶化，人类生存受到严重威胁。

We stand before an unsettling truth: greenhouse gas emissions have fueled global warming, triggered more frequent extreme weather, and accelerated environmental decline. This ongoing deterioration casts a shadow over the very future of human survival.

我的极简衣橱

简约潮流，生态新衣

My Minimalism Wardrobe

Simple Style, Eco-Friendly Fashion

衣橱是我们探索绿色低碳生活的起点。

构建一个风格独特、自我自信的时尚殿堂。在这里，不再有过度消费的喧嚣，每一件衣物的选择都经过了深思熟虑，它们是极简衣橱的主力。

简约才是潮流。这并非对时尚的割舍，而是对过度装饰的告别。在这个衣橱，我们深深感受到简单设计的力量，和那种专注本质、摒弃多余而产生的美。

拥抱生态新衣的理念，选择环保、再生、可持续的材质。在衣物的选材、制作过程中，力求减少对环境的影响，品味简单中的精致，感受时尚的深度。

Our journey toward a sustainable lifestyle begins with our wardrobe—a space where a unique style and confident self-expression come together. When every garment is thoughtfully selected, the noise of overconsumption fades away, and a minimalist wardrobe takes shape.

Simplicity is not a sacrifice of fashion but a farewell to excess. It unveils the beauty that emerges when we focus on what truly matters and letting go of the unnecessary.

Embrace eco-friendly fashion by choosing recycled, sustainable materials that are gentle on the environment. In every step, from selection to creation, we strive to lessen our impact on the environment, finding grace in simplicity and mindful choices.

爱上棉麻制品

踏入棉田，轻抚柔软的棉花，感受轻盈的自然气息。
棉麻制品亲肤透气、柔软舒适。用棉麻的朴素美学，唤醒对材料与生活的感知。

Discover the Charm of Cotton and Linen

Explore a cotton field, feel the softness of the cotton, and connect with nature. Cotton and linen products are gentle on the skin, breathable, and delightfully soft. Their simple aesthetics inspires a deep appreciation for nature's beauty.

镜前的低碳智慧

镜前的选择融入节俭理念，蕴含低碳智慧。
衣服无需太多，通过巧妙搭配，尽显独特风采。

Sustainable Style Reflected in the Mirror

Let your wardrobe choices reflect both simplicity and sustainability. You don't need a closet full of clothes—just thoughtful combinations to express your unique style while embracing a low-carbon lifestyle.

拒绝“快时尚”

选择耐穿经典的款式，少买“时尚”却廉价、穿几次就扔的衣服。

Say No to Fast Fashion

When shopping for clothes, choose durable and classic styles over low-quality, trendy pieces that are often discarded after just a few wears.

衣物集中洗

衣物攒够一桶再去洗。减少洗衣机的使用次数，节水又节电。

Wash Clothes in Batches

Gather enough clothes for a full load before washing. Reducing wash frequency helps conserve both water and electricity.

少量衣物用手洗

放下手机，适度劳动，感受皂的芬芳。体会生活，心情也变得愉快起来。

Hand Wash Small Garments

Set aside your phone, breathe in the fragrance of the soap, and wash small garments by hand. Engaging in moderate work can calm your mind and lift your spirits

减少熨烫

少选需要熨烫的衣服，省时又省电。

Minimize Ironing

Choose wrinkle-resistant fabrics to reduce the need for ironing, saving time and energy in your everyday routine.

适量使用洗涤剂

过量的洗涤剂使衣物漂洗变得困难，浪费更多的水，增加化学物质排放，加重对水体和土壤的污染。

Use Laundry Detergent in Moderation

Adding too much detergent makes rinsing harder and wastes water. Use it sparingly to reduce chemical waste and protect our water and soil from pollution.

拒绝皮草

每一件皮草的背后都有一个血腥的故事。
拒绝皮草，时尚不需要以生命为代价。

Say No to Fur

Behind every fur coat lies a tale of bloodshed. Say no to fur—fashion should never come at the expense of life.

自然晾晒 阳光味道

与使用烘干机相比，自然晾晒更环保。
衣物在晾晒过程中吸收阳光的味道，让我们身心也仿佛沐浴阳光，快乐放松。

Air Drying: The Scent of Sunshine

Air drying is far more eco-friendly than using a dryer. As clothes hang in the sun, they absorb the radiance of sunlight, leaving us feeling as though we, too, are bathed in its warmth.

旧衣演绎新时尚

旧衣改造是对环保、时尚和个性穿搭的探索。通过巧妙的设计和改造，将旧衣演绎成时尚新品。

Old Clothes, New Fashion

Upcycling used clothes is an innovative way to explore sustainability, fashion, and individuality. With a touch of creativity and redesign, old garments can be transformed into stylish, modern pieces.

捐赠旧衣

把不需要的衣服洗净叠好，将它们送到需要的地方，是一种告别过往岁月的美好方式。

Donate Old Clothes

Wash, fold, and donate the clothes you no longer need. Giving them away is a meaningful way to bid farewell to the past and help others in need.

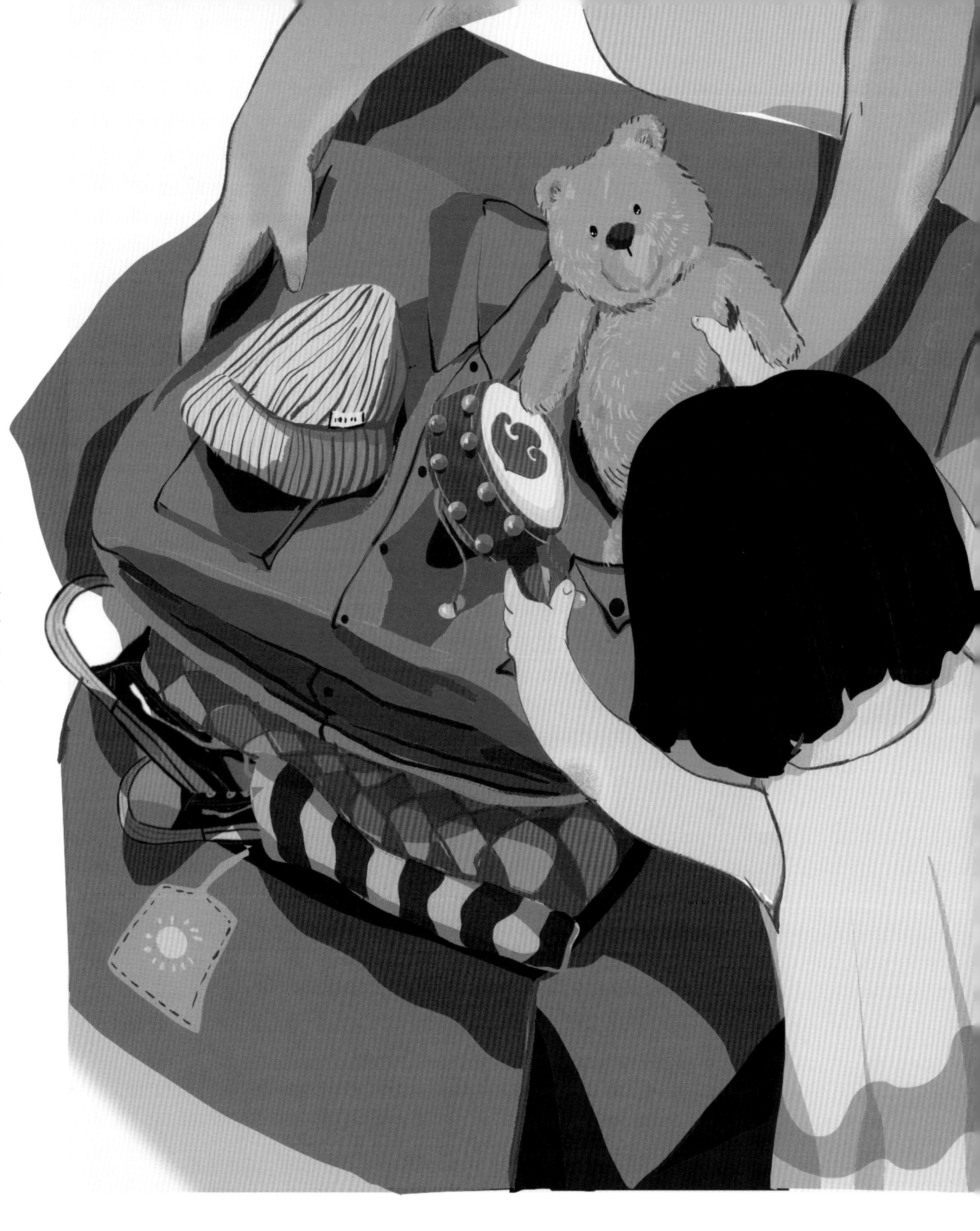

衣服传递爱

弟弟穿哥哥穿过的衣服，是低碳生活方式，也是爱的传递。

Threads of Love

Passing clothes from one sibling to another is a tender expression of love and a commitment to sustainable living.

15

我的人间烟火

简单厨艺，低碳美味

My Culinary Haven

Masterful Cooking, Sustainable Flavors

餐饮是我们探索绿色低碳生活的重点。

在一个简单而温馨的厨房里，探寻温暖岁月的人间烟火，感受食材的生命力，那份朴实与美味在每一道菜肴中得以体现。烹调不再是繁琐的任务，而是对生活的热爱。

追求低碳美味，注重食材的选择和烹饪方式的环保，选择本地、有机的食材，减少环境的负担。

崇尚低碳美味，每一道菜品都是对地球的爱护，是可持续生活方式的实践。

Cooking lies at the heart of our journey toward a low-carbon lifestyle.

In the kitchen, we find joy in the simple act of preparing meals, savoring the freshness of each ingredient. Here, simplicity and flavor come together, turning every dish into a small celebration of life.

Embracing eco-friendly practices can be as effortless as selecting local, organic produce. Craft each meal with the intention of honoring both the ingredients and the Earth. Let every dish be a mindful step toward a future of sustainability and harmony.

爱惜粮食

每一粒粮食都来之不易，
从小教育孩子不要剩饭。
吃多少盛多少。

Cherish Your Food

Every grain of food is precious. Teach children from a young age to avoid wasting food—only take as much as you can finish.

吃不了兜着走

美味不浪费，别任由餐馆倒进垃圾桶，将剩余餐食打包带走。

Leftovers to Go

Don't let delicious food go to waste. Instead of leaving leftovers behind at restaurants, pack them up and take them home to enjoy later.

不暴饮暴食

告别大吃大喝、挥霍浪费的用餐模式，让每一餐都成为对资源负责、对健康负责，文明而美好的时光。

Practice Mindful Eating

Resist the temptation of overeating and wasteful dining habits. Make each meal an opportunity to act responsibly with resources, a commitment to health, and a celebration of mindful living.

减塑捡塑

一次性餐具在我们的周围堆积，让地球慢慢窒息。减少使用一次性塑料制品。
捡拾身边的塑料垃圾，清洁环境。

Tackle Plastic Pollution

Single-use plastic utensils are piling up, slowly suffocating the planet. Make it a habit to reduce your use of disposable plastics and actively recycle plastic wastes to protect the environment.

减碳新“食”尚

自带工作午餐，减少点外卖次数，健康又环保。

Lunch Boxes: The Trendy Carbon Cutter

Bringing your own lunch to work reduces reliance on takeout and single-use, non-degradable utensils. Make a positive impact with each meal.

扫码点餐

选择扫码点餐、手机支付，开具电子小票和电子发票，就餐全过程无纸化，减少碳排放。

Scan to Order

Order via QR codes, pay with your phone, and opt for electronic receipts. A completely paperless dining process significantly reduces carbon emissions.

电气厨房

提倡用电磁炉代替燃气灶，更加清洁、便宜。无明火设计降低火灾隐患。
智能厨电让烹饪更便捷高效、清洁安全，引领现代生活新风尚。

Electric Kitchen

Switch from gas stoves to electric cooktops—they're cleaner, more affordable, and, most importantly, their flameless design reduces fire hazards. Smart kitchen appliances make cooking easier, more efficient, and safer, setting a new standard for modern living.

不在户外烧烤

户外烧烤会造成大气污染，还可能破坏绿地、引发火灾。

Rethink Outdoor Cooking

Outdoor barbecues, while popular, contribute to air pollution, damage green spaces, and in some cases, increase the risk of wildfires.

光瓶行动

水是宝贵的资源，每生产一瓶水都会产生碳排放。随手带走没喝完的瓶装水。

Empty Your Bottle

Consider the environmental cost of bottled water. Always take your unfinished bottled water with you.

我的个性水杯

少喝瓶装水就是减少垃圾。随身带个自我标签十足的水杯，做个环保达人，挺酷的！

Personalize Your Hydration

Drinking less bottled water means creating less waste. Carry a reusable water bottle that reflects your personality and be an eco-conscious trend setter!

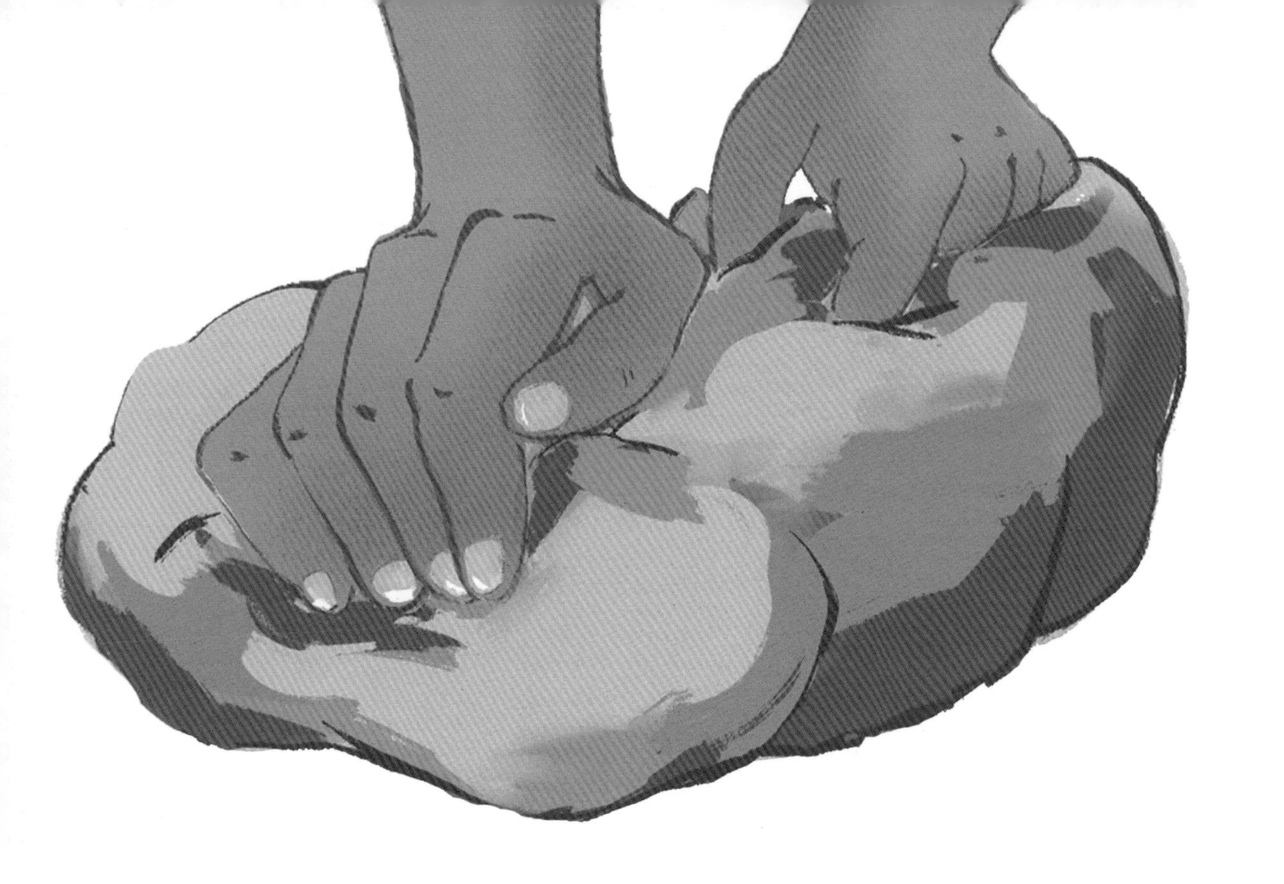

蔬果当绿植

把蔬果当绿植，好看又好吃。

Edible Decor: Vegetables and fruits as Greenery

Incorporate vegetables or fruits into your home décor—they're not just beautiful, they're delicious too!

买当季蔬果，离自然更近

与大自然的钟摆同频，优先选择充满天然能量、美味又低碳的当季蔬果。

Choose In-Season Produce

Align with nature by choosing in-season fruits and vegetables. Seasonal produce is tastier, more nutritious, and has a lower carbon footprint.

我家附近有早市

清早，蔬菜水果早市开张，只要一个小推车，在家附近就能准备好一天所需食材。
绿色生活从早市开始，方便又新鲜。

Morning Farmers' Markets

Explore the morning markets near your home for fresh, local produce. Shopping locally supports community farmers and helps reduce carbon emissions from transportation, packaging, and storage.

自备环保袋

大量一次性购物袋方便了自己，麻烦了地球。
选择可重复使用的购物袋，做绿色低碳生活代言人。

Bring Your Own Reusable Bag

Disposable plastic bags may be convenient, but they're a burden on the planet. Replacing them with reusable bags is a simple change that can make a profound difference.

冰箱不堆太满

冰箱东西放太多，不利于冷空气循环，影响食品保鲜，还会加大能耗。冰箱的秩序就是生活的秩序。

Don’t Overfill the Refrigerator

Overloading the refrigerator disrupts air circulation, reduces food freshness, and increases energy consumption. A well-organized fridge reflects a well-organized life.

提前解冻食品

提前用冷藏室解冻食品，不临时用微波炉解冻，省时又节能。

Defrost Food in Advance

Thaw food in the refrigerator ahead of time instead of using the microwave at the last minute. It saves both time and energy.

我的居家日记

轻松生活，清新品味

My Home Journal

Sustainable Living, Refreshing Aura

居家天地是我们绿色低碳生活的秘境。

轻松生活、追求舒适并非为了奢华，而是为了内心的安宁，在宁静中做好自我。

每一天起居，每一处布置，每一次购物，都是生活态度的表达；

每一天用电，每一次洗涤，每一次垃圾处理，都是环保理念的外现；

每一个盛物小竹篮，每一个开满鲜花的阳台，都是绿色低碳生活的品味。

从充满清新与生机的家出发，生命一路绿意盎然。

Home is where our green, low-carbon lifestyle quietly thrives.

Living simply and seeking comfort isn't about luxury—it's about finding inner peace and creating a space that nourishes the soul.

Every daily routine, every thoughtful arrangement, and every purchase is a reflection of our way of life.

Every time we conserve energy, wash clothes, or sort waste, we put our sustainable values into practice.

Each small bamboo basket, each balcony in bloom, reflects the essence of a mindful lifestyle.

From this haven of tranquility, we move forward, carrying the gentle whispers of nature in our daily steps.

我家是零碳微单元

我家屋顶有个小型光伏“发电厂”，可以满足全家用电需求。

A Home Powered by Green Energy

Harness the power of solar energy to significantly reduce your reliance on fossil fuels. A rooftop solar power system supplies clean electricity for your home’s lighting, appliances, and even electric vehicles.

我家轻装修

轻装修，重装饰。家庭装修不追求复杂的风格，注重简约和实用，节省成本、空间更开阔。

Simple Renovation, Stylish Decoration

Enhance your home with small, impactful changes instead of grand renovations. Choose meaningful, functional decor that breathes simplicity into your space. This approach opens up your space while keeping costs down.

住被动式建筑

我家住被动式建筑，充分利用自然通风和太阳能等可再生能源，以及优良的保温和隔热材料，降低房屋对能源的消耗。

The Benefits of Passive Housing

A passive house takes full advantage of natural ventilation and solar power to minimize energy consumption. High-quality insulation and heat-resistant materials help lower utility bills and reduces your environmental footprint.

以竹代塑

推进以竹代塑，从源头减少塑料垃圾产生。全球塑料污染治理可以从快速生长的竹子这里寻找答案。

Replace Plastic with Bamboo

Using bamboo as a substitute for plastic could significantly reduce waste at its source. The key to solving global plastic pollution may lie in bamboo, a fast-growing and sustainable alternative.

用啤酒当叶面肥

过期啤酒不要倒掉，用洁净的软布蘸啤酒，轻轻擦拭绿植叶片。擦灰尘的同时，还可以进行叶面施肥。叶片变得肥厚有质感，更翠绿、更有光泽。

Use Beer as a Leaf Fertilizer

Don’t let that unfinished beer go to waste! Dip a clean, soft cloth into the beer and gently wipe down the leaves of your plants. This beer-infused wipe nourishes the leaves while removing dust. The result? Thicker, more vibrant leaves, glowing with a rich, radiant green.

洗脸巾再用一回

用过的洗脸巾是不是随手丢掉？且慢，别扔，还能再用一回。洗一下，拿它擦家具、擦地板，效果都不错。

Reuse Disposable Face Towels

Do you throw away your disposable face towel after just one use? Wait—don't toss it yet! Wash it and reuse it to wipe down furniture or clean the floor.

少用一次性纸杯

生产一次性纸杯需要消耗树木资源。有些一次性纸杯内部覆有一层塑料薄膜，需要几十年才能降解。
塑料薄膜还会向热饮中释放微塑料颗粒，危害人体健康。

Use Fewer Disposable Cups

Producing disposable paper cups requires cutting down countless trees. Many of these cups contain plastic linings that are non-biodegradable and can release microplastics into hot drinks.

多用自然光

合理设计窗户和采光布局，让自然光成为白天室内的主要照明来源。书桌尽量靠近窗户，白天多用自然光。

Maximize Natural Light

Design your living spaces to make natural light the primary source of illumination during daylight hours. Position work and living areas near windows to take full advantage of daylight.

空调适温

建议夏天空调温度设置在26 ~ 28℃，节能又健康。
人体在这个温度下比较舒适。
温度过低，室内外温差过大，影响人体健康。

Optimal Air Conditioning Settings

Set your air conditioner between 26℃ and 28℃ in the summer to balance comfort with energy efficiency. This temperature range prevents excessive cooling and reduces the health risks associated with drastic indoor-outdoor temperature differences.

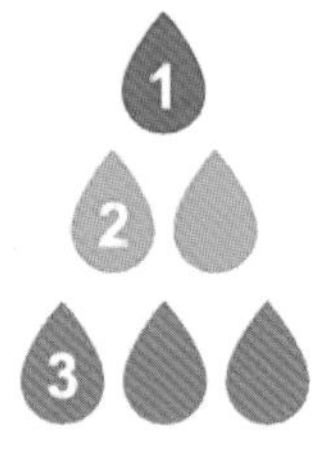

关注水效标识

中国水效标识是表示用水产品的水效等级的信息标签。
关注水效标识，选择节水产品。

Pay Attention to Water Efficiency Labels

Water efficiency labels provide key information about a product's water-saving capabilities. Prioritize products that demonstrate low water consumption to conserve this vital resource.

能效高 1

2

中　等 3

4

能效低 5

关注能效标识

中国能效标识是表示产品能源效率等级等性能指标的信息标签。
关注能效标识，选择节能产品。节约下来的能源最绿色。

Understand Energy Efficiency Ratings

Pay close attention to energy efficiency labels when purchasing appliances. Choose higher-efficiency products to save more energy—the greenest energy is the energy saved.

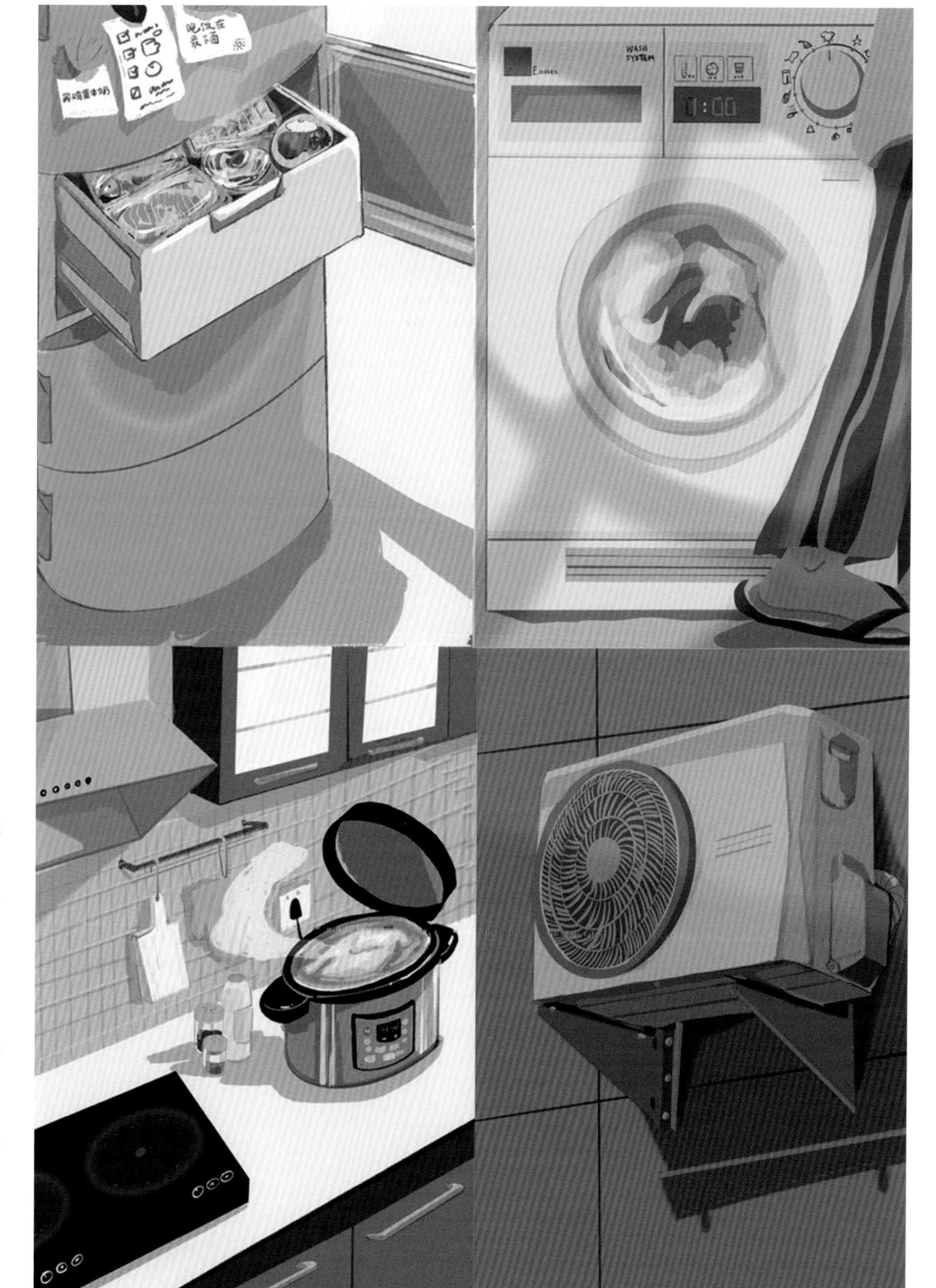

避免长时间冲淋

搓澡时关闭喷头，避免长时间冲淋。使用新型节水喷头可以节约水资源，将预热水收集起来洗衣拖地，也是节水好办法。

Shorten Shower Time

Turn off the showerhead while scrubbing, take shorter showers, and use modern water-saving showerheads to conserve water. Collect the preheated water for other household tasks like laundry and cleaning.

电热水器更低碳

电热水器没有明火，不排放二氧化碳，相比燃气热水器更清洁、更低碳。

Choose Electric Water Heaters

Electric water heaters operate without open flames and release zero carbon dioxide, making them a cleaner, more eco-friendly alternative to gas models.

不开着电视睡觉

困了就上床休息。开着电视睡觉既浪费能源又影响睡眠质量。

Don’t Sleep with the TV On

When drowsiness sets in, head straight to bed instead of drifting off to the glow of the screen. Falling asleep with the TV on wastes energy and disrupts the quality of your rest.

声控小夜灯

选择声控小夜灯，方便节能又温馨。轻声吩咐，自动开关，不用在黑暗中摸索。

Voice-Activated Night Lights

Voice-activated night lights are convenient and energy-efficient. They switch on or off with simple voice commands, providing illumination only when needed.

电器待机也耗电

手机充完电，记得断开插座电源；空调、电视机、机顶盒等长期不用，记得拔掉插头或关闭插座上的电源开关。

Electronics Consume Power in Standby Mode

Unplug your phone charger once it's fully charged. Disconnect electronics such as air conditioner and TV during extended periods of inactivity to prevent unnecessary power consumption.

出门前“三关”

出门前，例行检查水、电、燃气开关。节约能源，防止跑水、失火。

Routine Safety Checks

Make it a habit to check that the electricity, water, and gas are turned off before leaving home. This routine helps conserve resources while preventing water leaks or fire hazards.

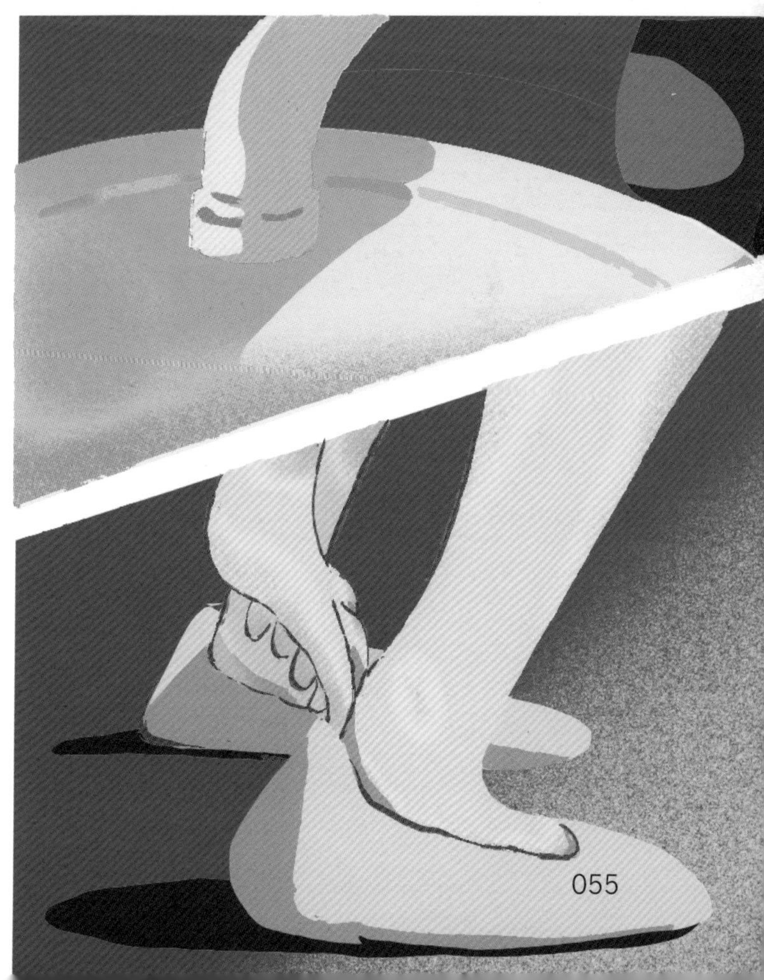

使用洗碗机 干净又省水

洗碗机在密闭空间内循环用水，利用喷淋臂上的喷孔冲刷餐具，比手洗更干净、更省水。

Use a Dishwasher

Modern dishwashers maximize water efficiency by recirculating it in a sealed environment, with precision nozzles that thoroughly clean every dish. They are more hygienic and conserve more water than handwashing.

节水龙头更划算

使用节水龙头，不滴不漏，开关方便、迅速。

Install Water-Saving Faucets

Upgrade to water-saving faucets that reduce water flow and minimize waste without sacrificing performance. These faucets conserve water over time, making them a cost-effective solution in the long run.

使用 LED 灯

LED 灯比白炽灯、日光灯更节能、使用寿命更长、光线更健康。使用 LED 灯是低碳生活的一部分。

Investing in LED Lights

LED lights are significantly more energy-efficient and have a longer lifespan compared to incandescent and fluorescent bulbs. Choosing LED lighting is a simple yet impactful step toward a more sustainable lifestyle.

不可“放任自流”

刷牙、洗脸、涂洗手液的时候，不要一直开着水龙头。

Don’t Let the Water Run

Don’t leave the faucet running while brushing your teeth, washing your face, or using soap. Every drop of water is precious—don’t let it go to waste.

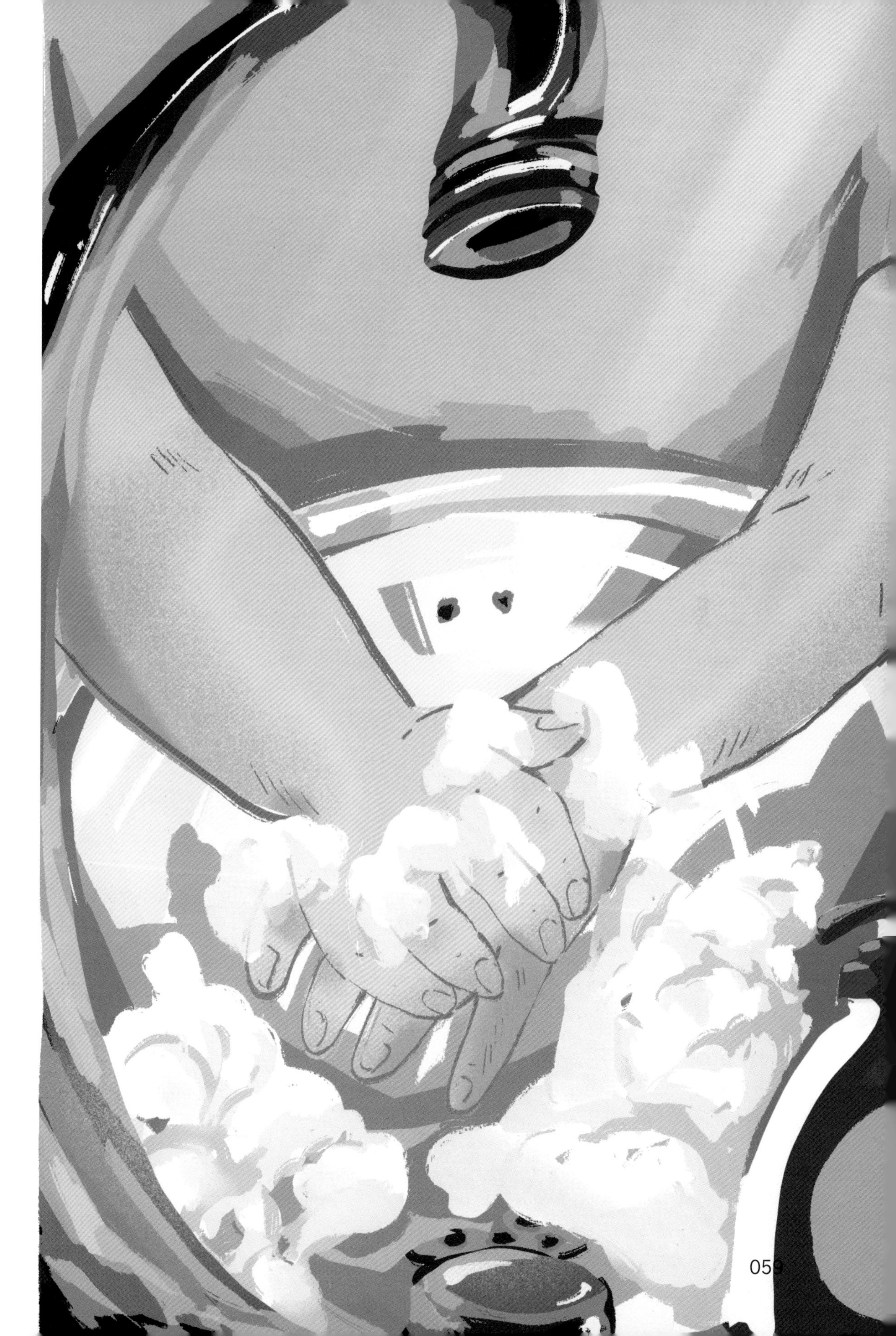

废物巧利用

用过的瓶瓶罐罐并非垃圾，通过巧妙改造，可以变成花盆、花瓶或其他有趣的装饰品。

Repurpose Materials Creatively

Used bottles and jars aren’t just trash. With a bit of creativity, they can be transformed into flowerpots, vases, or other unique decorations.

住高层也可以亲近自然

植物是给家庭提气的“软装”，生机勃勃是植物的状态，也是家人的心情。

Bringing Nature into High-Rise Living

Incorporate plants into your home decor to bring a touch of nature indoors. Plants uplift spirits, improve air quality, and contribute to a vibrant living environment.

我家花开　城市花开

无需大片土地，自家阳台就可以营造一个缤纷空间。
阳光透过花叶缝隙洒下，空气中弥漫着花香，我家花开，城市花开。

A Blooming Balcony Brightens the City

You don't need a big yard—a balcony is enough to create your own vibrant garden. A carefully decorated balcony garden can fill your apartment with the sweet scent of flowers. As your balcony blossoms, so does your city.

厨余垃圾变肥料

厨房里的菜叶、果皮、剩菜、剩饭等厨余垃圾，经过发酵腐熟，可以变废为宝，成为花花草草的优质有机肥。

Turn Kitchen Waste into Garden Gold

Kitchen scraps like vegetable leaves, fruit peels, and leftover food can be transformed into rich organic fertilizer through composting. Start a compost bin and turn your waste into a resource that enriches your garden soil.

养鱼的水可以用来浇花

鱼缸里换下来的水中含有多种营养物质，用来浇花再好不过。不仅节约用水，而且花儿长得更好。

Nourish Plants with Fish Tank Water

Fish tank water is rich in nutrients, making it an ideal, natural fertilizer for houseplants. By reusing this water, you not only reduce waste but also provide your plants with the vitality they need to flourish.

花花草草喝上再生水

将再生水用于厕所冲洗、园林灌溉、洗车或城市道路清洗等地方，节约宝贵的水资源。

Use Reclaimed Water

While reclaimed water isn’t suitable for drinking, it can be put to good use in other ways. Wherever possible, use it to flush toilets, water gardens, and clean outdoor spaces.

拒绝过度包装

层层包装，华而不实。包装体积大、实际产品小，喧宾夺主。
提倡简约设计，拒绝购买过度包装的商品。

Avoid Excessive Packaging

Flashy, oversized packaging often offers little substance and overshadows the product itself. When shopping, choose items with minimal packaging to help reduce waste.

就地回收包装盒

包装盒不要一扔了之，交给商场或快递小哥就地回收，可以循环利用，减少废弃物。

Recycle Packaging Locally

Don't just toss out those used packaging boxes! Instead, pass them on to a nearby shop or your local courier for reuse. With packaging always in demand, your boxes can easily find a second life.

我的出行指南

简单足迹，环保印记

My Eco-Friendly Journeys

Light Footprints, Green Paths

出行是低碳生活的“落脚点”。

少开私家车，首选公共交通，选择低碳或无碳排放的出行方式，降低交通成本，提高城市出行整体效率。

倡导“135”绿色出行方式，即 1 公里内步行，3 公里内骑自行车，5 公里左右乘坐公共交通。

让我们每次出行都留下绿色脚印。

Travel is where our low-carbon lifestyle finds its path.

Drive less, prioritize public transport, and choose low or zero-emission options whenever you can.

Adopt the “135” philosophy: walk for journeys under 1 kilometer, ride a bike for those up to 3 kilometers, and take public transit for distances around 5 kilometers.

Make every journey a stride toward a more sustainable world.

短途步行

步行是一种简单又经济的有氧运动，每天快走 30 分钟，特别适合忙碌的上班族。
说走就走，徒步“读”城，给身心更多的快乐和自由。

Take Short Walks

Walking is a simple, economical way to stay active. Make it a daily habit to walk for 30 minutes, especially if you’re a busy professional. Step outside, wander through your city, and rediscover the joy of reconnecting with your surroundings.

666路

骑共享单车

绿色达人，零碳出行， 扫码骑车，乘风快意。

Ride a Shared Bike

Scan with your phone, hop on, and enjoy the freedom of a carefree ride. Riding a shared bike is a sustainable way to travel, leaving no carbon footprint behind.

“4+2”出行

自驾游时可以带上自行车，4 轮 +2 轮出行，在合适的路段来一场惬意骑行。

“4+2” Travel

Bring along your bike on a road trip—four wheels plus two. When you reach your destination, switch to the two wheels and enjoy a serene ride through fresh landscapes.

乘坐公共交通

合适的距离，尽量乘坐公共交通。
停下匆忙的脚步，在车厢放空自己，欣赏窗外的风景。

Take Public Transportation

Whenever you can, choose public transit. Step back from the hurry of life, let your mind wander, and enjoy the scenery unfolding beyond the window.

不把后备厢当仓库

汽车后备厢长期堆放重物会显著增加油耗，尽量给汽车“减负”。

Don’t Turn Your Trunk into a Storage Unit

Extra weight in the trunk increases a car’s fuel consumption. Lighten the load to boost your car’s fuel efficiency.

燃油车选小排量

小排量汽车耗能少，排放的二氧化碳和空气污染物也少。选择小排量汽车可以改善城市的大气状况。

Choose Fuel-Efficient Vehicles

Cars with smaller engine displacements consume less fuel and emit lower levels of carbon dioxide and other pollutants. Choosing a car with a smaller engine can help create a healthier urban environment.

“元小源”
北京绿色生活平台

人人有个“碳账本”

注册个人碳账户，记录自己的碳足迹。衣食住行，每个生活细节都记录于“碳账本”。

Track Your Carbon Footprint

Sign up for a personal carbon ledger to monitor your carbon footprint. From commuting to dining, understanding the environmental impact of your daily activities can encourage more sustainable choices.

不在家洗车

现代化洗车房的水循环系统可以减少水的消耗和洗涤剂对环境的污染。

Avoid Washing Your Car at Home

Modern car wash facilities use water recycling systems that greatly reduce water usage and prevent detergent-filled runoff from polluting the environment.

购买新能源汽车

新能源汽车不烧油，可以大幅减少二氧化碳等温室气体排放，减少一氧化碳等污染物排放。

Switch to Electric Vehicles

Electric vehicles don't rely on fossil fuels, drastically reducing emissions of carbon dioxide and other harmful pollutants like carbon monoxide. It's a cleaner, more sustainable choice for the future.

低碳飞行

乘坐使用可持续航空燃料（生物质燃料）的低碳航班。

Low-Carbon Flying

Choose flights powered by sustainable aviation fuels, such as biofuels, to lighten your carbon footprint in the air.

乘坐高铁更低碳

高铁依靠电力驱动。与飞机相比，乘坐高铁大大减少了碳排放量，是一种低碳的长途出行选择。

High-Speed Rail Over Flying

Powered by electricity, high-speed trains have a significantly lower footprint than air travel. It is an excellent choice for medium-distance journeys.

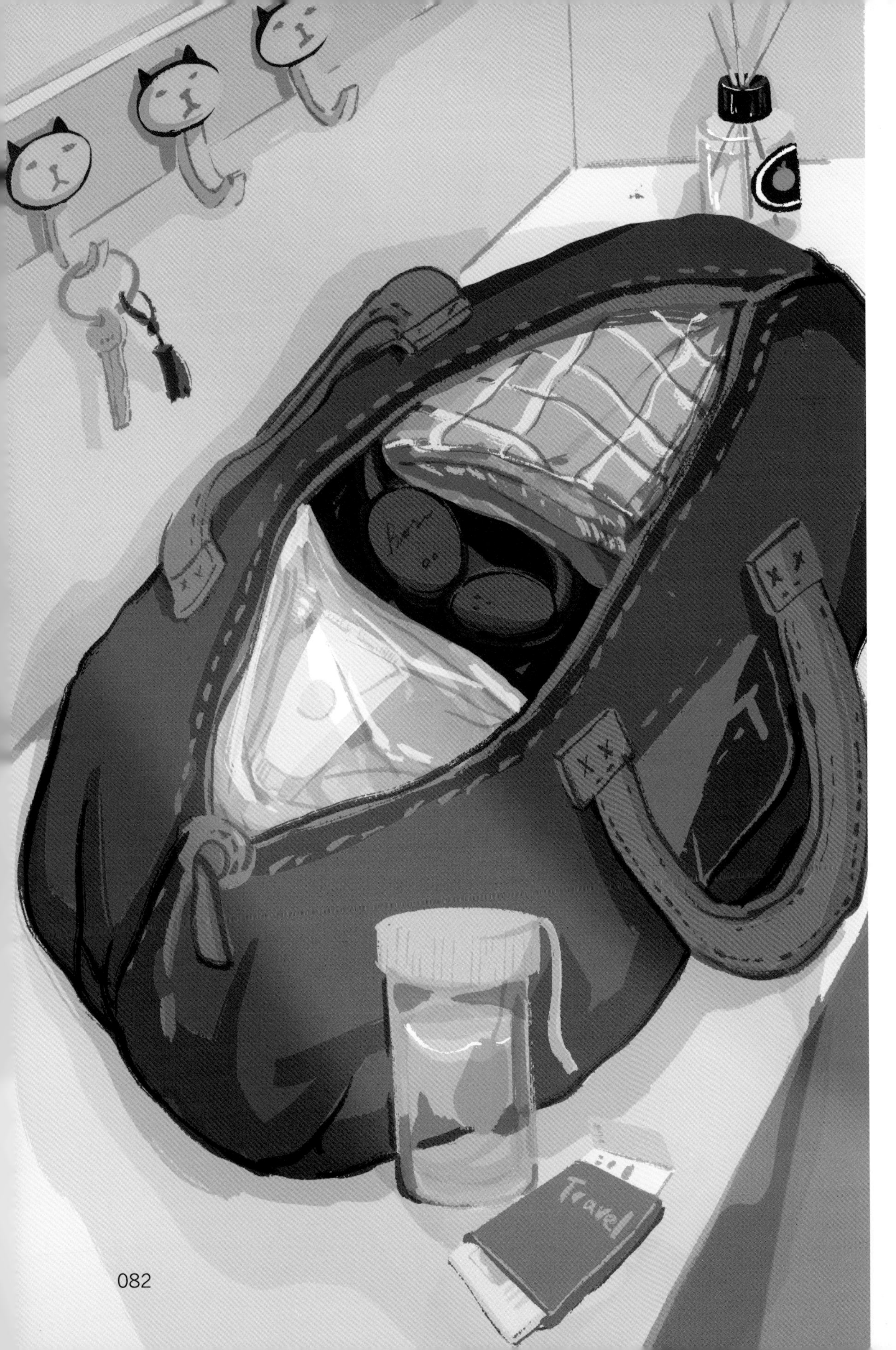

住酒店自带洗漱用品

自备洗漱用品，减少一次性用品的使用。更环保、更卫生。

Bring Your Own Toiletries

Carry your own toiletries when staying at hotels to reduce disposable waste. It's a simple step that makes your travels more sustainable.

线上支付

购买水电、办理银行业务、支付账单，通过线上支付，减少了交通出行。

Go Digital with Payments

Manage banking, pay for utilities, and settle bills online to reduce the need for travel, which in turn lowers your carbon footprint.

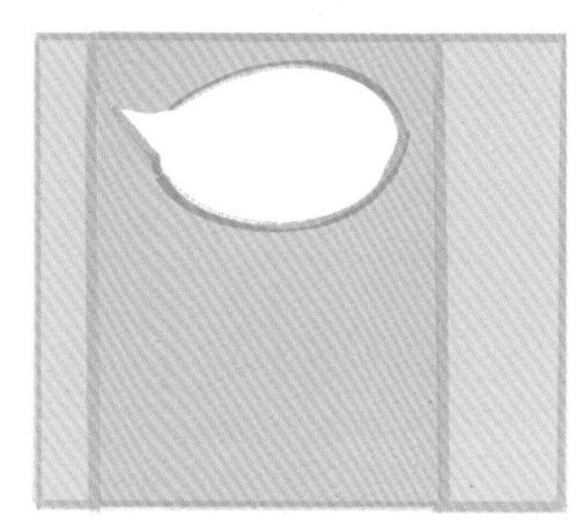

乘电梯等一下

关闭电梯前等一下赶过来的人。等待几秒钟，减少一次电梯运行。节能减碳，与人方便。

Elevator Courtesy

Take a moment to hold the elevator for others rushing to catch it. A few extra seconds can save an additional trip, conserving energy and reducing carbon emissions. Plus, it's a simple act of kindness!

勤走楼梯

走楼梯是一项有氧运动。低楼层住户少乘电梯、多走楼梯，有益健康。

Take the Stairs More Often

Consider taking the stairs as a simple aerobic exercise. If you live on lower floors, use the stairs instead of waiting for the elevator to save time and sneak in a bit of extra movement to your day.

坚持每周户外运动

走出健身房，坚持每周一次以上户外运动，释放生活和工作的压力。

置身于大自然，在青山绿水间呼吸新鲜空气，心情格外舒畅。

Weekly Outdoor Activities

Make outdoor activities a weekly habit and aim to spend more time outdoors overall. Stepping outside reduces sedentary indoor time, fills your lungs with fresh air, and helps you attune to the rhythms of nature.

享受森林浴

远离喧嚣的城市，走进森林，打开视觉、听觉、嗅觉，让身心完全沉浸在天然氧吧中，在绿荫中减压，来一次洗涤身心的森林浴。

Forest-Bathing

Escape the hustle and bustle of the city and unwind in nature. Immerse yourself in a natural, oxygen-rich environment that rejuvenates both body and mind. Let the greenery soothe your mind, ease your stress, and deepen your connection with nature.

我的办公学习

朴素空间，绿色理念

My Sticky Notes

Little Reminders, Big Impact

上班族也是绿色低碳生活的主力军。

开始办公了，碳足迹就在脚下延伸，无纸化办公、线上会议、双面打印、节约耗材，这些加分；浪费纸张、开长明灯、电脑长期待机，必须扣分。

办公空间深藏绿色品味，绿植装点、绿意盈盈，每一天都充满了生机和力量。

办公空间彰显绿色品质。注重环保工作方式，减轻环境压力。

每一天有工作成果，每一天有绿色成果。

Office workers are key players in the low-carbon movement.

With every workday, our carbon footprint grows. Actions like going paperless, holding virtual meetings, printing double-sided, and conserving supplies add to our green score. But wasting paper, keeping lights on unnecessarily, or leaving computers in standby mode will subtract from it.

A truly green workspace carries its own charm—plants in every corner bring a sense of freshness and vitality to each day. It's a space that reflects our commitment to the environment, where sustainable work habits lighten the strain on our planet.

Every day brings professional progress, and with it, small victories for the environment.

重复使用回形针

除工作必需，多用可重复使用的回形针、长尾夹等，少用订书针和胶水。

Reuse Paper Clips

For everyday office tasks, opt for reusable paper clips and binder clips instead of staples and glue whenever you can.

COLORED
PAPER CLIPS
PORJEC
ROPOSA

无纸化办公

使用数字化全流程管理，信息传递更加高效。
大幅减少纸张等耗材，节约资源、降低成本。

Go Paperless

Shift to a fully digital workflow for efficient information transfer. Going paperless conserves resources and cuts operational costs by minimizing the use of paper and other materials.

线上会议和线下会议相结合，在实现高效沟通的同时，减少不必要的出行。

Virtual Meetings Reduce Carbon Footprint

Combine online and in-person meetings to maintain effective communication while reducing unnecessary travel. This hybrid approach not only lowers the carbon footprint from commuting but also supports a more flexible work environment.

双面打印

多采用电子文档代替纸质文件，减少打印次数，尽量选择双面打印。

Double-Sided Printing

Whenever possible, prioritize digital documents over paper ones. When necessary, opt for double-sided printing to minimize paper use.

办公室绿植

办公室的每个角落都有绿色点缀，繁忙的工作中打开一扇通往大自然的窗。

Greenery in the Office

Incorporate plants throughout the office to create a more inviting and visually appealing work environment. Greenery improves air quality and provides a moment of calm to the daily grind.

下班断电

下班前，断开插座电源，避免电脑、打印机、饮水机等设备长时间待机。

Power Down After Work

Before leaving the office, unplug power sources to prevent computers, printers, water dispensers, and other devices from staying on standby for long hours. This simple habit helps avoid unnecessary energy consumption and lowers electricity bills.

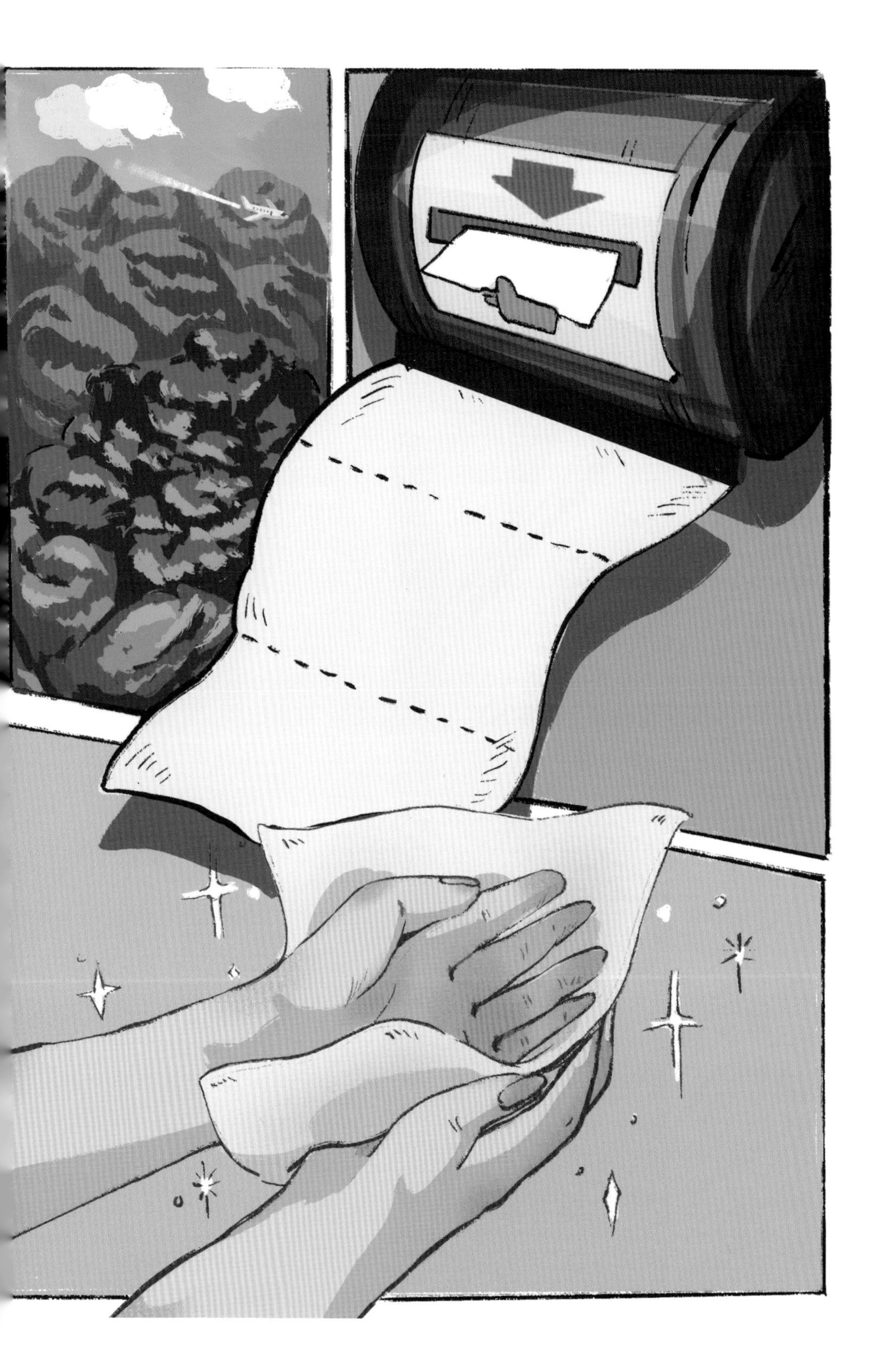

只用一张纸

洗完手，轻轻甩一甩，只用一张纸，保护一片绿。

Only One, Please

After washing your hands, give them a gentle shake and use only one paper towel. This small habit can make a big impact by conserving resources.

用纸少留空

一张纸没写几笔就丢掉很浪费，养成尽量写满的好习惯。

Fill Scrap Paper Completely

Make full use of scrap paper and write on both sides of the paper before recycling it.

更换笔芯

常用的水笔笔芯是可更换的，墨水用完，更换笔芯即可，不要连笔一起扔掉。

Refill, Don't Replace

Most ballpoint pens have replaceable ink refills. When the ink runs out, simply swap in a new refill instead of discarding the entire pen.

用旧报纸练书法

旧报纸不要扔，可以拿来练书法。

Repurpose Old Newspapers

Give old newspapers a second life by using them for calligraphy practice instead of discarding them.

电子阅读

使用电子阅读器，减少纸质出版物。节约纸张，保护森林。

Switch to eBooks

Invest in an e-reader and embrace digital reading. Reading online reduces the demand for paper books and magazines, saving paper and helping to preserve our forests.

享受亲子阅读时光

亲子阅读时间，是家庭宝贵的陪伴时刻。远离手机、设置静音，捧起书来，以低碳的方式关注和陪伴孩子，享受美好温馨的时光。

Parent-Child Reading

Parent-child reading time is a moment of shared wonder. Set aside your phone, open a book, and dive into a story together. It's an eco-friendly way to bond with your child and create memories that linger long after the last page is turned.

捐赠书籍　分享智慧

清晨的阳光洒在脸上，山区的孩子捧起一本书，坐在窗前静静地阅读。
捐赠书籍，分享智慧，让隽永书香弥漫在每一个孩子心间。

Donate Books, Share Wisdom

As the morning sunlight gently kisses their faces, children from impoverished regions are exploring worlds through donated books—opportunities they might never have without your kindness.
A book donation is more than a gift—it's a way to share knowledge and imagination, planting seeds that can bloom in every child's heart.

我的绿色地球

和谐乐章，万物共生

My Planet Earth

Nature's Symphony, Life in Harmony

我们只有一个地球，我们共有一个家园。

尊重自然、顺应自然、保护自然，我们每一天都在做环保拼图。

珍爱地球、天人合一、万物共生，我们每一天都在寻觅绿色意义。

清新的绿、醉人的绿、蓬勃的绿、诗意的绿，绿色地球就在我们的行动中。

We have but one Earth—a sanctuary we all share.

With every act of respect and each gesture of care, we place another piece in the fragile puzzle of preservation.

Cherish our Earth, strive for harmony between humanity and nature—envision a world where all living things flourish together.

Green is fresh, green is enchanting, green is vibrant, green is poetry—our Earth renews itself with every mindful step we take.

保护城市绿地

绿地是城市之“肺”。保护绿地，爱护植被，不攀折花木，不挖野菜。

Protect Urban Green Spaces

Green spaces are the lungs of the city. Cherish and safeguard them—avoid picking flowers, breaking branches, or digging up wild herbs.

仰望星空

环境改善了，城里的孩子也能看到满天繁星。辽阔无垠的宇宙让我们更加敬畏自然。

Gazing at the Stars

As the environment heals, city children can once again marvel at a sky filled with stars. The vast, boundless universe inspires a profound reverence for nature, reminding us of our humble place within its grand design.

在爱的逻辑中成长

宠物是孩子们天然的朋友。让孩子们从小亲近自然，养成保护动物的观念，在爱的逻辑中成长。

Growing Up with Compassion

Pets are children's natural companions. Encourage them to bond with animals from an early age, instilling the values of love, care, and respect for all living creatures.

垃圾精细分类

混放成垃圾，分类变资源。

Sort Waste Properly

Mixed waste is just trash, but with proper sorting, it can become a resource.

垃圾精准投放

按照垃圾桶上的标识投放不同品类的垃圾。

Dispose Waste Precisely

Place your trash in the correct bins according to the labels. Avoid tossing kitchen waste into recycling bins.

做环保志愿者

看到路边或草丛里的废纸、瓶盖、包装袋，弯腰随手捡。做环保志愿者，从身边的小事做起。

Volunteer for the Environment

When you spot litter, such as paper scraps, bottle caps, or wrappers along the roadside or in the grass, take a moment to pick them up. Becoming an environmental volunteer begins with these small, everyday actions.

闲置物品交易

“你的旧物我需要”。网上闲置物品交易让旧物派上新用场。

Trade Unused Items

“One man’s trash is another man’s treasure.” Use online platforms to trade your unused items, giving them a chance to be valuable and purposeful in someone else’s hands.

无痕出游

外出游玩，随手带走垃圾，不带走一草一木，尽量保持景区原貌。

No-Trace Travel

When exploring the outdoors, always take your trash with you. Leave the natural landscape as you found it.

全民义务植树

让植树成为风尚，每年种树三五棵，献给地球一片绿。

National Tree

Planting Initiative—Plant three to five trees each year to help reforest our planet and restore its natural beauty.

云端植树

通过捐款实现“云端”植树和保育、修复森林，每个公民都能参与到绿化事业中来。

Digital Tree Planting

Donate online to support tree planting and forest conservation efforts. Every citizen can play a part in building a greener future for our planet.

拒绝过度捕捞

过度捕捞导致海洋生物数量减少，使一些物种濒临灭绝，导致生物多样性丧失，影响整个海洋系统的生态平衡。

Advocate Against Overfishing

Overfishing depletes marine life, pushes species to the edge of extinction, and disrupts the delicate balance of the ocean's ecosystems. Support sustainable fishing practices to protect marine biodiversity and preserve the health of our oceans for future generations.

清理海洋垃圾

塑料污染威胁海洋生物的安全。海龟、海鸟、鲸鱼等海洋动物可能因塑料垃圾堵塞消化道而死亡。

Clean Up the Ocean

Plastic pollution poses a grave threat to marine life. Sea turtles, seabirds, and whales often suffer and die when plastic waste clogs their digestive systems.

没有买卖就没有伤害

不要为了人类的贪欲而伤害野生动物。从源头治理，禁止盗猎野生动物，禁止非法买卖野生动物制品。

No Trade, No Harm

Human greed should never be an excuse to harm wildlife. Report poaching and the illegal trade of wildlife products to help protect vulnerable species.

保护动物就是保护我们自己

让我们一同深思，在人类与动物共同生存的地球上，生态平衡之于人类福祉的重要性。
动物是人类的朋友，保护动物就是保护我们自己。

Protecting Animals Is Protecting Ourselves

Let us take a moment to reflect on humanity's vital role in maintaining ecological balance. Animals are our companions on this shared Earth; by safeguarding their well-being, we, in turn, ensure our own.

保护生物多样性

丰富多彩的生物世界为人类提供了美学享受和文化灵感。保护珍稀濒危物种，保护生物多样性，世界因共生而美好。

Protect Biodiversity

The rich tapestry of life enriches us with beauty, inspiration, and cultural depth. By protecting endangered species and preserving biodiversity, we ensure a more vibrant and enduring planet—one in which future generations can cherish and flourish.

关注人类可持续发展

在极地，冰川和海冰正加速消融，仿佛大自然在默默流泪。我们开始自觉减碳，就是小心翼翼地拼一张可持续发展地图。

Focus on Sustainable Development

As glaciers and sea ice melt away in the polar regions, it's as if nature weeps in silence. By consciously reducing our carbon footprint, we can forge a path toward a more sustainable future.

绿色地球　生生不息

草长莺飞、鹤舞大地。绿色地球 ，生生不息。
朝着我们向往的方向，插上坚强的翅膀，我心飞翔。

Green Earth, Everlasting Life

Where grass rises and birds sing, where cranes dance across the earth's this is the pulse of a vibrant Earth, ever-thriving. Guided by our vision, let us soar together toward a shared horizon.

I would like to extend my heartfelt thanks to all the friends who have supported and contributed to the writing and publication of this picture book. A conversation with Mr. Yang Kun three years ago sparked the idea of promoting a green, low-carbon lifestyle in an approachable and engaging style. Mr. Zhu Hongren consistently encouraged me to broaden my horizons and adopt a global perspective. Mr. Du Shaozhong organized the expert review team for this book and personally guided the revisions, while Mr. Zhou Jianping, who served as the chief reviewer of my previous work *Electric Power of China*, continued to provide meticulous guidance.

Special thanks to Mr. Du Xiangwan, Mr. Pan Jiahua, Mr. Shu Yinbiao, Mr. Jiang Yi, Mr.Sun Baoguo, Mr. Liu Jian, Mr. Zhang Laiwu, Mr. Qiu Baoxing, Mr. Fu Chengyu,Mr. Zhu Hongren, Mr. Yang Kun, Mr. Yan Shidong, Mr. Tu Ruihe, Mr. Du Shaozhong, Ms. Wu Qimin, Mr. Xu Jintao, Mr. Liu Guanzhong, Mr. Liu Suwen, Mr. Zhou Jianping, Ms.Tao Lan and ,Mr. Li Changshuan who served as academic advisors for this book. Their encouragement and support have been a great source of strength.

I am grateful to Ms. Chen Huiling, whose artistic touch and unique sense of color brought warmth and joy to the illustrations. A special acknowledgment goes to Mr. Lian Zijian, whose translation work opened this book to a global audience. His insightful suggestions and attention to detail enriched the content, ensuring that the essence of the work resonated with readers around the world. I am thankful to Deputy Director Ding Li of the State Grid's External Liaison Department for her interest in this book, to Deputy General Manager Yan Jun of Yingda Media Investment Group for his support, and to Deputy Director Cao Rong and Editor Yang Yang of China Electric Power Press for their dedication.

Additionally, my sincere thanks go out to the many friends and colleagues who assisted with this project, including Lü Kun, Zhang Jidong, Li Qiang, Wang Bo, Miao Rong, Yuan Xueguo, Wang Shitao, Yue Xiaodong,Wang Zhongyuan, Wang Dong, Chen Binhui, Chen Zisen, Wang Mozhu, Wei Qiuli, Fu Changchao, Yang Zeying, Lü Jing, Han Yaling, Huang Wen, Ding Huilan, Cui Xiyuan, Guo Yicheng, Zhou Zhixing, Xie Yiming, Lin Fei, Liu Jiawei, Du Bohan, Chen Haoyang, Liu Tianqi, and Xue Heqing.

To all of you, my deepest thanks!

致谢

在此，我要向支持和帮助本书编写和出版的所有朋友表示最诚挚的感谢。

三年前与杨昆先生的一次谈话，我深受启发，产生了以轻松活泼的风格普及绿色低碳生活方式的想法；朱宏任先生一直鼓励我不断精进，以更加开阔的视野走向世界；杜少中先生为本书组织了审稿专家团队，并亲自指导书稿修改；周建平先生是拙著《大国电力》的总审稿人，他一如既往地帮助本书精心打磨。

感谢杜祥琬先生、潘家华先生、舒印彪先生、江亿先生、孙宝国先生、刘坚先生、张来武先生、仇保兴先生、傅成玉先生、朱宏任先生、杨昆先生、闫世东先生、涂瑞和先生、杜少中先生、吴绮敏女士、徐晋涛先生、柳冠中先生、刘素文先生、周建平先生、陶岚女士、李长栓先生担任本书学术顾问，他们的支持让我信心倍增。

感谢绘图作者陈慧玲女士，她对色彩的独到把握让绘本更加温暖欢快。感谢英文译者廉子健先生，他从不同视角对内容提出修改意见，他的翻译让本书有机会面向全球读者。感谢国家电网对外联络部丁莉副主任对本书的关注，感谢英大传媒集团晏俊副总经理的支持，感谢中国电力出版社曹荣副主任和杨扬编辑的付出。给本书提供帮助的朋友和同事还有吕昆、张继栋、李强、王博、缪荣、袁学国、王仕涛、岳晓东、王忠源、王东、陈斌惠、陈自森、王墨竹、魏秋利、付长超、杨泽英、吕竞、韩亚玲、黄文、丁惠兰、崔曦元、郭弈成、周志行、谢一鸣、林非、刘佳伟、杜伯涵、陈昊阳、刘天奇、薛禾卿等，在此一并表示感谢！

2024 年 9 月

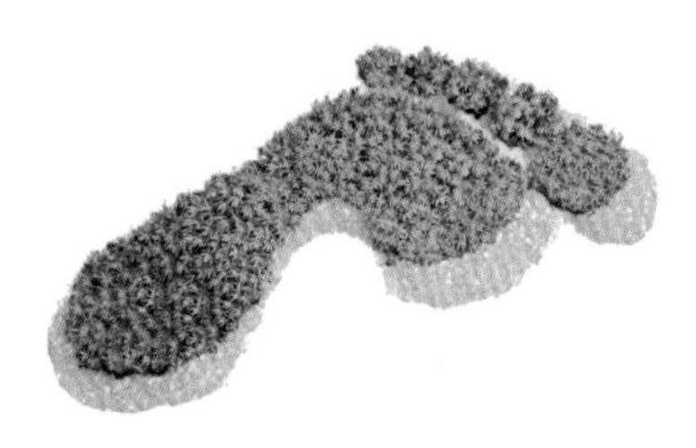

始于心　践于行

Born in the Heart, Manifested in Action